PERSPECTIVES
WITH
PURPOSE

PRINCIPLES OF PURCHASING &
SUPPLY CHAIN LEADERSHIP

PETER BOWIE DILL

LEADERS ACROSS THE GLOBAL SUPPLY CHAIN COMMUNITY AGREE,

"PERSPECTIVES WITH PURPOSE"

UNLOCKS THE HIDDEN VALUE BETWEEN PURCHASING

AND OTHER SHARED SERVICES,

WITH THE ADDED BONUS OF AI INSIGHTS

"Whether you are new to the field or assuming your first leadership role, *Perspectives with Purpose*, is chock-full of sage, experienced-based advice on how to navigate complex internal and external environments and relationships. And, as adoption of artificial intelligence becomes as significant to business processes as was the introduction of the internet in an earlier era, Peter Dill reminds us that effective human relationships are the ultimate source of successful business outcomes. Dill makes a convincing case that trust, clear communication, cross-functional consultation and alignment, and developing a 360-degree view of issues and conflicts are essential skills in supply chain."

-Tom Derry, CEO, **Institute for Supply Management (ISM)**

"Peter's Cut-to-the-Chase focus on Balanced Perspective is the reminder we all needed to break through the ingrained silo approach and better manage the matrix to deliver value at a whole new level!!!"

-Jeff Martin, Director, Global PC Service Supply Chain, **HP**

"*Perspectives with Purpose* pushes beyond analytics and processes to illustrate that successfully managing complex relationships is the true foundation of an effective supply chain leader and organization. Peter gives succinct, practical examples that are useful for a new supply chain professional or a seasoned leader. *Perspectives with Purpose* is a great playbook to harness the internal corporate cross-functional alignment that is required to navigate global supply chain risk and opportunities."

-Adrian Bregnard, General Manager Supply Chain, **Shell**

"A masterclass in distilled wisdom. Perspectives with Purpose takes decades of global supply chain experience and focuses it into actionable leadership insights. Whether you're navigating the daily routine of business or dismantling corporate silos, this book provides the 'guardrails of purpose' necessary for any leader to excel. Bravo!"

-Jim Guinn II, Senior Cybersecurity Leader **&** *Advisor*
Former Senior Managing Director, **Accenture**
Former Senior Partner **Ernst & Young, LLP**

"As C-Suite leaders look to Supply Chain to help their organizations navigate through economic headwinds, Perspectives with Purpose is the perfect primer to harness cross functional alignment."

-Xenophon Koufteros, Professor Department of Information & Operations Management
Fellow of the Decision Sciences Institute, Jenna & Gavin R. Guest Professor in Business Administration, Eppright University Professorship in Undergraduate Teaching Excellence, Editor in Chief-Decision Sciences Journal Mays Business School **Texas A & M University**

"In this era of fluctuating tariffs, geopolitical uncertainty, and isolationism, Peter brings refreshing new Perspectives which will help leaders refocus on delivering supply chain sustainability"

*-Achim Heyne, Senior Consultant, **ATH Consulting, GERMANY***

"Peter Dill has done it again!

In *Perspectives with Purpose*, he delivers concise, invaluable nuggets of wisdom drawn from decades of supply chain expertise. Whether you're a seasoned professional or new to the field, you'll gain practical insights to elevate your performance and advance your career. Dill masterfully emphasizes the power of open, positive communication across functions—unlocking innovation and transformative improvements. Blending timeless core skills with forward-looking perspectives on emerging technologies like AI, this book is an essential resource for every supply chain leader.

Read it, enjoy it, and grow!"

*-John C. Ploetz, Retired Director & President, **ExxonMobil Catalyst Services, Inc.***

"Peter Dill shares his wisdom on a sometimes-overlooked success factor of Supply Chain leadership - proactively developing positive partnerships with other disciplines within your own company. While Supply Chain is generally acknowledged to own the company's external relationships, Perspectives' focuses on the unique (and, BTW, existential) opportunity that all Supply Chain leaders have - to unite the 'home team' around shared goals. Peter's many frank and illuminating personal examples make this an accessible and engaging must-read for all company stakeholders".

-Lisa Haley, Supply Chain executive (retired)

To my Lord and Savior Jesus Christ:

Your perspective is all that matters.

To my wife, Leslie:

The encouragement behind all my successes…
sharing them with you is Everything.

And my children: Travis, Jessica, and Luke:

I am proud of who you are,
and what you will become.

PERSPECTIVES
WITH
PURPOSE

PRINCIPLES OF PURCHASING &
SUPPLY CHAIN LEADERSHIP

PREFACE

Based on the success of my first book, Control what you Can, I was invited to share these concepts at national conferences across a wide range of sectors including Supply Chain, Logistics, Oil and Gas, Information Technology, and others. This also led to repeat invitations to major universities to help those majoring in Supply Chain better understand the concepts that would lead to success in their chosen career. The opportunity to engage at so many levels with a variety of leaders across several sectors was extremely energizing. The opportunity to inspire the next generation was indeed inspiring.

Students, executives, and other leaders appreciated the distilled (short), focused format, and made a point to compliment the unexpected **PERSPECTIVES** that applied to several of the topics addressed in the book. I have been blessed with a wide variety of roles across several sectors, along with living and working in several countries and 35+ years of experience. This has given me the opportunity to consider situations, projects, people, and corporate functions from many different angles. I received many questions regarding if, and when I might author another book, and the idea of sharing a different look at various functions from a Supply Chain/Purchasing perspective was born.

The title was conceived to align not only with my previous focus on Principles of Purchasing and Supply Chain Leadership, but also to include the guardrail of *Purpose.* The format of short, distilled chapters, followed by a focused benefit statement remains vital to give the reader clear, applicable takeaways. In the age of the new and evolving AI landscape, I have added a Perspective from this angle, as well, to provide further insight and application of these leadership principles.

The topics were selected for chapter headings and prioritized based on experiences and observations over recent years and months. These key perspectives were also prioritized in order based on where each could result in significantly improved value for the corporation.

Thus, *Perspectives with Purpose* came to life. Much of these reflections came from years in Supply Chain, and others were pulled from more recent overarching views of daily engagement in my current role. As with *Control what you Can*, some of this may be very familiar, and a helpful reminder of what we as Supply Chain leaders should harken back to as we lead our teams. My hope is that the reader will also be blessed with a few new insights when considering how to engage at a new level with other shared services. It is my wish that these insights will allow you to not only break down barriers and silos, but to deliver an exponentially new level of value to the enterprise you work for, and a new level of fulfillment in your role as a Supply Chain leader.

None of the creative content of this book was derived from Artificial Intelligence (AI)

TERMS / DEFINITIONS

Note that the industry standard definition of Supply Chain covers all the items bought by an organization including goods and services, transportation, warehousing, logistics and utilities etc. The term Purchasing is also used to describe this function. Typically, the term Procurement refers to the more tactical aspect of sending out relatively smaller bids and managing more complex and higher value contracts that have been previously negotiated by the Strategic Sourcing team. The terms Sourcing and Strategic Sourcing typically refer to higher level strategies and negotiations centered around Category Management. Different corporations may use the terms differently. At my current company, Supply Chain refers more to inventory management of direct material. At a major oil company Procurement is used to describe all Purchasing functions. In this book the terms Supply Chain and Purchasing will be used interchangeably. Category Leaders are responsible for developing high level strategies and negotiating high level contracts for a specific type of good or service (furniture, MRO, contingent workers, lighting, metals) that a corporation needs to purchase. The terms vendor and supplier will also be used to interchangeably represent entities that provide goods and services to customers represented by Supply Chain/Purchasing.

Chapter Summary

1. Negotiation, Timing, Relationships & Tariffs

2. Supply Chain/Purchasing & Expectations

3. Legal & Dollars

4. Finance/Accounting & Sponsorship

5. HR/Leadership & Credibility

6. Engineering & Curiosity

7. Supplier Quality & The Triad

8. Data Analytics & Pick up the Wrench

9. Purchasing Operations & Supply Chain support

10. Site & Balanced Perspective

11. Change Management & The Crisis

12. Perspectives with Purpose & Legacy

1

Negotiation Timing, Relationships & Tariffs

When I attend social gatherings, THAT question always comes up. What do you do? I have gotten into the habit of flipping the question. "Based on our conversation for the past 10 minutes and your observations while at this gathering, what do you think I do?" The rest of the conversation, on many occasions, has unfolded like this.

"Are you in sales?"

"No, but perhaps I should have been."

"Are you a Pastor?"

"No, but thank you very much, I am a Christian, and very proud of my Faith."

"I give up."

"I have been working in Supply Chain / Purchasing for 30 years or so."

"Well, you really don't seem like the Purchasing type and certainly aren't like most Purchasing people I have dealt with...but I bet you are really good at negotiating with suppliers."

The truth is 70% of the negotiation I have done over my 30+ year career is not with suppliers/vendors, but within my own company first, seeking alignment with the shared services we will discuss in this book. This is a key concept from the second chapter (One Voice) of my first book, Control what you Can. This concept aligned with the perspectives of Timing and Relationships are key to successful negotiations.

Timing

During a recent trip to Egypt, my wife, youngest son, and I decided to visit the pyramids, and we wanted to do it in style. We did our research and made arrangements to stay at the Mena House hotel in Giza, which was nearby with impressive views of these stunning monuments to the pharos. After long flights, and transferring to the hotel, the next morning we were ready for our adventure.

We made it known that we wanted to approach the pyramids of Giza via camelback. The guides nodded and we were ushered over to a spot where they ceremoniously offered us ghutrahs (red & white checkered headscarves) and wrapped our heads in them in the spirit of Lawrence of Arabia. When I attempted to negotiate for these accoutrements, much Arabic was spoken very quickly, but no money changed hands. Instead, we were led over to a waiting group of camels and introduced to all of them by name. Then we were instructed on how to climb onto the saddle of a sitting camel and how to hang on, as they first extended 4-foot rear legs, and hung on tightly, while our mounts rocked up on their front legs. We were all proudly settled 7 ft in the air atop our assigned humps…..and THAT is when our guide, comfortably standing on the ground, in full control of the camel ropes to my family, decided to open negotiations relating to how much this camel ride to the pyramids, including photo shoot, was going to cost. Needless to say, I did not have any leverage and had not paid any attention to the quietly evolving perspectives of Timing and Situational Awareness.

Happily, the ride and views were worth every piastre. The key purpose of sharing this perspective is to manage your surroundings, (sourcing strategy, vendors, stakeholders & business unit leaders) before they manage or place unnecessary limitations on you and your sourcing strategy and timing. As a Category Leader, it is your responsibility to drive your sourcing strategy, paying close attention and checking in regularly with your internal and external stakeholders to understand where reasonable adjustments need to be made. Silence from either may not necessarily mean alignment.

Relationships

Relationships are another key to effective success in negotiations. Let's now consider a different perspective on relationships and flipping the traditional negotiation script by first considering what your vendor needs to be successful...and seek ways to give it to them. Sound crazy?

Consider the perspective of an automotive component supplier who, after being on the receiving end of a one-way negotiation which resulted in such a hit to their profit margin, could no longer afford to do business with one of the Big Three auto manufacturers. This supplier then announced to the manufacturer that they were not only going to stop supplying key parts, but also retreated with the key tooling (big, engineered equipment) they owned. The financial impact of stopping the assembly line as well as the operational impact, was catastrophic. Rethinking this impact and perspective, why not endeavor to negotiate fairly with vendors, and build reliable relationships, which will lead to more reliable production schedules. The ultimate goal centers around negotiating value for your company while allowing your vendor to earn an acceptable profit margin.

When one considers negotiation, it is vital to first be aware of the nature of the relationship and related view that each partner has when engaging. Let's take this from a more personal perspective. Think of a person you may not get along with. Now replay in your mind, not from your mind's eye, but from this person's view, all the actions, inactions, and engagements you have had with this person. Did you only reach out to them when you needed something, or when they did something that offended you? How did you first meet? Was it because you made the effort to make a new friend or was it because there was a crisis and you were thrust into a harrowing situation and you two had to fix something. Introductions and personal background sharing were either secondary or rushed.

Do many of the questions above now leave you wishing you made different decisions relative to how you first engaged with this person? Would other decisions have resulted in a trusted friendship rather than a shifty adversary? Now you have a better idea of the value of Suppler Relationship Management (SRM), which can also easily be applied to internal customers or shared functions. Let's now apply this to professional business relationships. Think about the last time you received "that call" from a Vice President, identifying a very big issue that you had no idea how to fix. Further, you knew that fixing it would require cross-functional engagement with, yes, that person, function, or business unit leader with whom you have zero trust and a limited relationship. That relationship may be weak due to reasons in the paragraph above, or the effort was never made...or perhaps you burned that bridge.

We have all been there before, and you know that as soon as possible you need to schedule a cross-functional meeting to fix the situation. Obviously, the best time to get to know the leader of that other function or try to repair that relationship is *NOT* in the middle of a crisis. Put another way, it is exponentially harder to work through a crisis with someone you do not like, trust, or at least have a healthy business relationship.

Therefore, it pays dividends to sincerely make the effort to get to know those in other functions *before* that call from senior leadership. This is the difference between strategic partners and suppliers. There is mutual benefit in investing time to align strategic vision with strategic partners. This also applies to internal customers and shared services.

More importantly, it is the right thing to do and turns what could be a daily grind into a meaningful experience. Early engagement allows you to put the gifts you have been blessed with to the use for which they were intended and appreciate the same in others. Think of how much it means to you when another leader praises you for your ability to work with others. Make it a point to inject this into your network relationships.

It is notable from an international perspective that most negotiators in countries outside of the United States focus on developing a personal relationship first. This is before considering if a business relationship and a contract will ever be created. Why? Because the focus is on building TRUST first and only then, a business relationship. The advantage of this, when done in a professional and moral manner, is that when, not if, an issue arises, it is first viewed from a trust perspective. In many cases this may mean that instead of immediately escalating through leadership at both customer and supplier, and engaging legal teams, both parties take responsibility, engage in meaningful discussion, and resolve the issue. The ideal situation results when both sides focus on doing whatever it takes to maintain the long-term, trusted relationship.

One of my key collaborators who leads the supply chain function at a major global technology corporation heartily agrees with this sentiment, especially when working with Asian strategic partner suppliers.

When considering how AI can be applied to negotiations, there are already bots and agents which attempt to negotiate after preparing and issuing RFI's and RFPs and evaluating the responses. It will be up to Supply Chain leadership to determine how much of this negotiation and at what level this will be executed by Buyers and Category Leaders vs. AI agents. But either way, one should consider utilizing AI to create the best digital twin of the vendor/supplier. The concept would include loading every data point regarding previous negotiating history, current financial status, and any related geopolitical factors into a database that would represent this vendor and then engage the AI agent in negotiations. Even if the mock negotiation does not go well, you will learn valuable lessons prior to the real thing. And certainly, have back-up strategies prepared in case the engagement deteriorates.

Tariffs

Any book published in this new era of tariffs would not be complete without a perspective related to the foundations of this drastically expanded negotiation arena. We will briefly address the triggers, tipping points, pitfalls, and a few recommendations to navigate and avoid significant negative exposure.

Tariffs are triggered when material goods are transferred across borders and a defined tax is required to be paid to the government of an importing country, typically based on Harmonized Tariff Codes (HTC Codes). My first exposure to this concept came in the mid-1980s, after 9 years living as a dependent in Saudi Arabia, when it was time to move back to the United States. My older brother enjoyed riding offroad dirt bikes through the desert, and it was time to ship this motorcycle, along with all of our other household goods stateside. My father has always been creative, and when he found out what he would pay in duties to ship this toy that may or may not have been priority in the 40-foot shipping container, he did some research. Why? Because the most prepared wins! He found, with support of the HTC Codes, that the cost of shipping two wheels and a motor was significantly less than shipping a dirtbike, and the disassembled treasured toy made it back to Texas!

So, be sure your Supply Chain team has a solid knowledge of HTC codes. This can be done in smaller companies by doing your best to educate your Category Leaders/Buyers. However, based on my previous experience leading a team of SMEs trained in this area at a Freight Forwarder for several years, I strongly recommend considering the option of outsourcing this function to a Freight Forwarder. This move will allow your Category Leaders/Tactical Buyers to remain focused on their core competencies (described in further detail in Chapter 9). From a compliance perspective, this will most importantly avoid significant federal fines and being placed on a related watchlist but also minimize tariff costs through due diligence of correctly selected HTC codes through the application of creative, yet legal solutions.

Before considering which tariffs may apply, ensure your team verifies that a tariff should be paid at all. Were the goods already in the country before the tariff was established and before the goods were imported? If so, no tariff should be paid. Be sure to confirm that your company is the importer of record. If it is the supplier, the supplier is responsible for paying the tariff.

Ensure through your purchase to pay process that the amount of the tariff is listed on the invoice, so a reasonable path forward can be developed by the responsible buyer. When evaluating whether the tariff should be paid, several paths forward should be considered. These include:

1) Select another supplier who may not pass the tariff on to you as the customer.

2) Negotiate with the supplier to cover all or at least ½ of the tariff.

- Be mindful that although the supplier (importer of record) is legally bound to pay the tariff, this does not imply that the customer is obligated to pay ANY of this tariff. This is negotiable by the relevant tactical procurement buyer.

- Also pay attention to contract wording which may permit opening pricing discussions around factors beyond their control, including unforeseen Tariffs.

3) Select another local supplier where possible.

4) Select a supplier in another country which has a more favorable tariff regime.

Finally, when negotiating, it is reasonable to remind yourself and your counterpart that there are at least two perspectives in any negotiation. Your job as a professional Supply Chain/Purchasing leader is to represent and leverage to the best of your ability your corporation's position to deliver maximum value. However, always respect and understand the supplier's perspective.

Perspective with Purpose

Refocus on Timing, Situational awareness, and Relationships with functional leaders as foundations for value driven negotiations.

Keep a watchful eye on tariffs & consider alternatives.

AI Perspective

In most negotiations, it would be risky to rely solely on AI for sense of Timing/Situational Awareness/Relationships, but prompts centered around positive/negative case studies involving these concepts could be valuable.

2

Purchasing/Supply Chain

&

Expectations

Recently during a recurring meeting with Chief Procurement Officers (CPOs) across various business sectors, we explored several expectations the C-Suite had for Purchasing/Supply Chain. It became quite apparent that whether these leaders were responsible for billions of dollars of spend across Oil & Gas, Technology, Health Care, or Chemicals, one of their primary goals centered around meeting the needs of their business and site stakeholders. This priority was followed closely by having solid working relationships with their peers in shared services including Finance, Legal, Engineering, Logistics, & Supplier Quality. Meeting these priorities requires a deep **perspective** of what drives the leadership of these stakeholders and other shared services….and how they may perceive Purchasing/Supply Chain.

Leaders working in other shared services may perceive Purchasing/Supply Chain as a team of contract writers whose primary function is to "do the paperwork/contracts" once other functions, (ex. Engineering, Sales) have decided and told a supplier what they want and perhaps even how much they are willing to pay for it. This might even include sending out a bid once in a while. This may be standard practice in some companies who have not been exposed to the benefits of Category Management and/or Strategic Sourcing. If you work in a company where this is the norm, it will continue to be the case until you take the initiative and demonstrate the benefits of Assurance of Supply, Quality, Service, Price, and access to Innovation which come from industry standard supply chain methodologies.

As value driven, difference-makers, we should place expectations on ourselves, as C-Suite leaders do, to take the initiative to seek every opportunity to deliver value when we encounter operational issues, supplier deficiencies, and the like. Category Leaders and their reports must strive to become Subject Matter Experts (SMEs) in the following, at a minimum:

- Stakeholder needs, pain points, and related business unit strategies
- Category Spend overall and broken out by Sub-Category, Division, Site, Region
- Cost Drivers which affect the items above
- Market Research
- Innovations in the industry
- Geopolitical risk factors in the Category (ex. Tariffs, Supplier financial stress)

With this knowledge the Category Leader can educate Stakeholders at all levels of management across several functions, even when the best strategy may involve not so palatable aspects, including accepting some level of price increase or not going out to bid.

Many consider internet search aided by AI as a primary method to gather market intelligence. Let's consider including a different and more personal perspective when conducting market research to both understand a Category/Product from both the product and negotiation angle. Recently, when I started to research the purchase of a new vehicle, I did the obligatory online research of which manufacturers had vehicles in a certain sector, and the features of interest

(My Clear Requirements). Then I started to call dealers around the country to find the particular make and model loaded with these key features. I discovered interesting facts that the manufacturer was not advertising…like a significant recall for which there was yet to be a solution. I also stumbled on a salesman, who was also an SME on this vehicle, who was more than open to share which vehicles he would recommend based on his intense interest and ownership of the same vehicle. Later, another salesperson from another state, who, once he realized that it would not make economic sense for him to locate and sell me this vehicle, was more than willing to share several other very useful perspectives including:

- Why it was not economical for him to sell me this vehicle
- Rough cost of this vehicle with my favorite features
- A reasonable estimate of what it should cost to load this vehicle on an unenclosed car carrier to my location…and same for an enclosed car carrier.
- Which features could be found in another trim level package, which may make the total cost significantly less than what I originally had in mind

This reminded me of a key truth that is overlooked in many cases when conducting market research…a source/SME that does not have a stake in your negotiation and has expert knowledge of your product or service…**is invaluable**. This is why I seek a contact within my supply chain network who has purchased a particular good or service and have a deep conversation with them regarding the performance of this vendor. I certainly do NOT ask about pricing or other related details which would be unethical, but focus on general performance.

A fundamental expectation both supply chain leadership and your stakeholders have of Purchasing professionals is centered around cost estimation. My first year as a buyer was full of learning opportunities, and my first manager constantly stretched my thinking in a variety of ways. Some of these exposures were punctuated by the positive example below.

My manager would stop by my cube and observe that look of deep thought on my face and ask what was up. I shared that I was puzzled yet again about how I was expected to know what a benchmark price of a particular part should be. An example could be a connector for a set of wires to a headlamp. He would say something like, "well, what would you pay for a plug in at a major home improvement store?" I would respond with a figure, and he would suggest I start there, and suggest I then do the following:

1) Talk to our engineer to understand the differences between a household plug and the clear requirements for this automotive plug.

- If there is a Cost Engineering function, consult with this group.

2) Check on the price of what other plugs elsewhere on this vehicle platform cost, or what they cost on last year's model.

3) Check other sources of market research, including AI options.

4) Most importantly, ask the supplier to provide a Cost Breakdown of the cost they were proposing.

5) Then, certainly issue a bid to competitively validate the cost.

There are many books focused on the idea of developing cost models, but these basic steps are a reminder of a few methods to approach and negotiate cost.

Perspective with Purpose

Seek sources of Market Research within your network, from those who do not have a stake in your negotiation, which will provide trusted information.

AI Perspective

Interrogate AI as well for sources of Market Research, but be cognizant of the boundaries, including internal firewalls, which may require paid access and/or subscriptions to category specific information.

3

LEGAL

&

DOLLARS

Early in my career, and based on binge-watching various courtroom drama series, my perception of the lawyers that made up our corporate legal department was skewed drastically towards courtroom drama and litigation. You know the scene where a well-dressed lawyer presents opening and closing arguments with all kinds of flourish, poignant pauses, and glares at opposing counsel.

But at its core, Legal engagement is about the money. Take a look at a publicly traded corporation's annual report. In many cases it may include results of a multimillion-dollar lawsuit…either in favor/or against the corporation…which in some cases can significantly affect overall earnings. With this in mind, the role of the legal department is to protect the corporation through advising the Purchasing Category Leader regarding terms and conditions, associated liability, and liquidated damages. Certainly, the Category Leader is responsible for the commercial terms of the contract, but in many cases, the commercial terms are intertwined and influenced by the Legal team's interpretation of the terms and conditions.

For the reasons above, I no longer send my contracts to my very valued legal counterparts just for redlines. Instead, I make an effort to build trusting relationships with them, involve them early…to explain my Clear Requirements regarding Intent, and what is vital to make this deal happen. Based on my knowledge and negotiation with my sales counterpart, I absolutely EXPECT and encourage my sales counterpart to do the same with his/her legal team. It is not productive for him/her not inform me that they "sent it to legal" (as though their legal department was another company). I diplomatically remind my sales counterpart that their commission depends on communicating their

Intent and what is needed to close this deal. When both the Category Leader and the Vendor counterpart align on Clear Requirements, which are very closely linked to Intent of the contract...this renewed perspective can bring about a win-win productive engagement. This approach minimizes the amount of back-and-forth red lines by opposing counsel.

Industry standard best practice is to ensure that suppliers document acceptance of the purchasing company's standard terms and conditions as a condition of receiving the bid. Why? Imagine a scenario where a supplier is informed, they have met the quality, service, price, and technical requirements of the bid, and will be awarded the business, pending acceptance of terms and conditions. The supplier knows they have won the business. They have all the leverage and incentive to push back on terms and conditions that could be costly or simply not aligned with their business processes. This also results in significant waste of time negotiating further with the supplier...as well as internally from the purchasing perspective with stakeholders, legal, and other shared services which did not anticipate need for further engagement. Now consider the best practice scenario where the supplier or group of suppliers must agree to standard terms and conditions BEFORE receiving the bid. Logic dictates that the supplier will price their bid to include the cost, risk, and other factors involved in delivering the good or service based on these terms and conditions. If your organization does not have this process in place, I cannot stress enough the efficiency gain achieved through focus on this related to Clear Requirements discussed in my first book, Control what you Can.

Another misconceived perception the business unit or site leadership may have regarding Purchasing is that they are a bottleneck or delay relating to finalizing contracts. For example, a smaller company is bought and integrated into a larger global corporation. Prior to this transaction, the smaller company leadership was nimble and could make decisions regarding suppliers, without vetting contracts and even without bids. It is not surprising then, when leadership at this smaller company becomes frustrated by the relative delay when Purchasing Category Leaders take the necessary time to:

1) Confirm Clear Requirements from the business or site

2) Ensure that the vendors are vetted and agree to corporate Terms and Conditions to qualify to bid on the business. Which may include the following:

1) Review the initial contract proposed by the vendor.
2) Respond with the standard corporate contract template.
3) Engage in legal negotiations, along with commercial terms.
4) Develop and issue a bid to qualified vendors who are part of the established set of Preferred Suppliers.
5) Select the winning vendor based on best combination of Quality, Service, Technology and Price.
   a. The most effective model includes a weighted rating that Supply Chain and Stakeholders align with early in the sourcing process.
6) Submit the final selected contract for Purchasing approval.
7) Document the contract in the corporate contract repository.

All these steps not only protect the corporation from costly lawsuits, but this procedure is typically the required, mandated process…. which, if not followed by the Category Leader, may result in termination. To avoid a misperception of Purchasing, the Category Leader should take the time to diplomatically educate the newly acquired site or business unit leadership on the Clear Requirements related to this process. Even more importantly, they should build a positive, value-focused relationship, sharing the reasoning behind WHY the process is in place.

The final decision remains with business unit leadership to accept and approve strategy and related spend with a particular supplier...because it is their budget. In too many instances there is a misconception that the overall corporate budget is owned by Purchasing. While Purchasing, like other shared services, is responsible for its own relatively small portion of the budget, the vast majority of the corporate budget is typically the responsibility of various business units. Purchasing/Supply Chain has the ultimate responsibility to own and manage the commercial relationship with suppliers. In most mature corporations, decisions relating to high level strategies which involve supplier selection and significant spend are typically developed and presented by Supply Chain/Purchasing leaders and approved by a combination of Supply Chain, Finance, Engineering, Legal, and Supplier Quality.

Perspective With Purpose

Develop a solid relationship with your Legal counterpart, to align on their primary role of protecting the corporation from Legal and ultimately, financial exposure.

AI Perspective

It is healthy to explore options related to automated contract review but be sure the Category Lead along with Legal representative conduct a final review before the document is routed for approval.

4

FINANCE/ACCOUNTING

&

SPONSORSHIP

Early in my Purchasing career, my perspective relating to Finance/Accounting was quite different. Even with a finance focused MBA, I did not want to engage with Finance types, since they knew much more than I did about all things Finance and Accounting....and might just call me out. So, I just steered clear of them. This was quite a flawed strategy since, at most corporations, Finance is the ultimate judge and validator of achieved hard dollar savings.

Within most mature Supply Chain (Purchasing) organizations, there is some version of a Sourcing

Table, or regular strategy approval meeting, chaired by Sourcing leadership which includes representation by business unit stakeholders, Engineering, Supplier Quality, and, you guessed it, Finance. Wouldn't it make sense to align with Finance long before one presents his/her initial sourcing strategy and resulting savings? I learned by observation and the hard way that the answer is a resounding YES. So, from that point on, I made it a point to understand what constituted savings at each company I have worked for, and more importantly, got to know the Finance leaders who would be validating my savings.

Consider this hypothetical perspective. Mary, the Chair of the Strategic Sourcing Review Board (SSRB), receives the agenda for the monthly meeting. She sees that Joe is coming forward with a significant savings strategy for approval. Two days before the meeting, she calls John, VP of Finance, and asks if he is aware of Joe's savings initiative, and if he is supportive and will approve. This conversation could go two ways. One, "Mary, this is the first I have heard of this, and from a quick look at the executive summary, I'm not clear how Joe could claim this level of savings." Two, "Mary, Joe has been working closely with my team, has taken a few tips on how to organize his strategy and savings, and not only will I be quick to sign off on this initiative, but want to recognize Joe at the SSRB for the solid work he has done on this one."

Obviously, we all want to position ourselves to walk through door number two. However, so often due to the range of other tasks associated with aligning/negotiating with other stakeholders, and certainly suppliers, we forget Finance until it is too late. Take the time to build a trusted relationship with your Finance counterpart and consult them often and in detail relating to specific initiatives. In this way you will be high fiving them at the Sourcing Table in front of your mutual bosses and setting the example for others.

Perspective with Purpose

Give Finance every opportunity to help sponsor your savings initiatives.

AI Perspective

As you develop your relationship with Finance/Accounting, seek to understand how they are applying AI…it may significantly influence your savings calculation.

5

HR / LEADERSHIP

&

CREDIBILITY

When asked to summarize the positive impact of military training on my leadership style and related approach to problem solving, I reflect on my days in the Corps of Cadets at Texas A & M University and later, as a Lieutenant leading a team of US Navy Salvage Divers in the Pacific. The training prepared officers for a variety of unpredictable situations "topside," hundreds of feet underwater, and in the recompression chamber after a dive.

On one occasion, while anchored off the island of Java near Bali, a seasoned diver surfaced after an uneventful dive and lost feeling in his legs. The dive team immediately transferred him into the recompression chamber and initiated treatment. The hyperbaric treatment session lasted much longer than our binder-based treatment dive tables prescribed, and I was tasked with contacting the Diving Medical Officer in Hawaii for further guidance. Meanwhile, the ship's crew had engaged the ship's diesel generator to pull up the bow anchor and then proceeded to do the same with the stern anchor, which weighed 3 tons and was roughly the size of a small car. Then the generator failed, and the ship started to drift towards the nearby reef. All of this happened while the rest of the dive team was working to stabilize the stricken diver. The bridge team, managing the course and fate of the ship, had little time to react and chose a risky but "best of worst outcome" decision to engage both of our massive propellers and drag the anchor out into deeper water. There it could be manually, but very slowly, cranked up and secured to the stern. The diver later recovered.

Why include this anecdote in a section relating to HR and management of people? To illustrate that managing people (and lives) is often about making tough decisions that are best for your team when none of the choices are clear, optimal or will necessarily result in a straightforward path to success.

The primary lesson I learned from my experience in the military that I have applied time and time again to managing people is this... Get the job done, regardless of the resources you may or may not have and put your people first.

Another lesson I have learned relates to the opposite of the old adage... "Manage by Exception" Most managers are typically overloaded with work, and significantly understaffed. Thus, we tend to focus on signals from leadership, sites, business units, or our reports which, in turn, lead us to act. However, after many instances of "Quiet" or no signal I have regretted not looking into this silence. Just because internal stakeholders are not raising an issue or escalating to your management does not mean all is well. Nor are your reports necessarily content when you do not hear them complain. The same applies to the customer/vendor relationship. The first signal that a vendor has an issue with a customer is failure to deliver the project/service/product on time.

What is the simple, yet often neglected, solution to avoiding catastrophes in all of the situations above? Regular engagement. One may refer to this by its Strategic Sourcing nomenclature as Supplier

Relationship Management (SRM) or simply maintaining a professional level of engagement. However, it comes down to setting up regular calls, meetings, or breaking bread to CHECK IN. Ask vendors during Quarterly Business Reviews (QBRs), or more often as needed, if everything is going according to plan. Are they getting paid on time, has their financial situation changed, etc. Do the same with your internal customers…and other shared function partners (HR, Legal, Accounting, Operations, etc.) Set up regular one on one meetings with each of your reports to ask if they are satisfied with their career path, if they need training, do they understand what is expected…do they like being on your team? If your manager only has reviews with you once or twice a year, request a meeting more often to understand if you are meeting expectations and how he/she perceives your leadership style.

The relationship between Purchasing and Human Resources is pivotal based on the anecdotes and concepts noted above. As a Purchasing leader, you must constantly strive to not only support & develop those you lead to deliver results, but hire the best to replace them when they are promoted or move on. Seek the counsel of your HR counterpart when outlining clear requirements for a given role, the salary in the marketplace, and seeking interview guidance. You will also absolutely need HR insight when working through challenging personnel issues. As with other shared services, the time to get to know your HR counterpart is NOT in the middle of a sensitive issue or crisis. By the way, this is the same part of the organization that manages <u>your</u> paycheck and potential promotion (-;

The perspective HR is starting to apply based on AI is seismic when it comes to talent acquisition. Based on conversations with Chief Procurement Officers, along with senior HR leadership, there is a wide-ranging continuum when it comes to the skillset sought after to support the organizational design of the next generation of Supply Chain shared service. One extreme seeks employees with primarily strategic and high-level problem-solving skills, enabled by AI…while the other extreme pursues those who still have the foundational experience doing and learning from fundamental tasks of the Purchasing craft, learned during their first role as a buyer. As with many continuums, the most effective may be somewhere in between, depending on the sector and culture of the organization.

Let's first consider the cutting-edge new hire who has fully embraced AI. He/she makes it a priority to identify repeatable tasks, quickly defines the basic steps of these tasks, and then employs AI to execute them. If results from these tasks are outliers, they then flag these for human evaluation and execution. This new hire does this to allow more time and focus on strategic tasking where more value can be delivered by application of creative thinking. This person may have thrived in university classes where the professor veered away from assignments that involved more hands-on learning and encouraged much more application of AI when creating content and even significant focus on utilization of AI to solve complex problems. The capstone projects were laid out with no limits on application of AI, since this is the most significant shift in work and personal life since the internet. Perhaps this is just the new hire with the *mindset* and skillset we should all be seeking, right?

Now let's consider the pragmatic new hire who has embraced a more hands-on learning approach. He/she first works through the details of repeatable tasks to fully understand which steps can be eliminated and then streamlines the process. This also allows significantly more time to focus on strategic tasking where more creative application of problem solving can be applied. He/she had that professor who required students, even in this new age of AI, to execute the majority of their content without this tool, because without a deep understanding of basic elements & steps of a process, even something as mundane as creating a Purchase Order or Engineering Change Order, they would not be equipped to improve a Source to Pay process. The professor also required the content of papers and presentations to be developed without the aid of AI to develop the creative skills that would lead to more efficient problem solving on many levels as this employee grew professionally. Who wouldn't want to make an offer to this new hire?

Based on my engagement with department heads of Supply Chain programs and professors at several major universities, the application of AI in the classroom is indeed a challenge. AI is taught and encouraged at various levels, but fundamental execution is valued and applied to instill the vital skill of problem solving. Not surprisingly, this thinking is applied to many career fields. While PO creation in Supply Chain is becoming increasingly automated, so is account reconciliation in Accounting, and creating a job description in HR. But most readers with 5-10 years of experience started their careers doing these things, learned from them, and apply what they learned as they manage others, and certainly as they problem solve on a more strategic level.

Recently I was leading a preparation meeting for a conference panel I was going to moderate, focused on what an agile, future ready Supply Chain organization would look like. Senior leaders from both Supply Chain and Academia agreed that a combination of both AI savvy and pragmatic hands-on problem solvers will be the best make up of these organizations. It could be a mix of new hires with a primarily AI focused mindset along with pragmatic hands-on executers, along with new hires that were a combination of both. The main message was that we still, and always will, need a good measure of pragmatic, hands-on experience, even as AI changes the workplace.

Perspective with Purpose

Develop and maintain your relationships with HR and your reports sincerely and consistently, these are the foundation of your daily engagements...and results.

AI Perspective

The skillsets HR is helping you bring to your organization will be heavily influenced by the seismic shift towards AI but should be weighed against the value of experience from trial and error executing fundamental tasks.

6

ENGINEERING

&

CURIOSITY

I learned much regarding how to interact with engineers in my first role as a buyer at a major automotive manufacturer many years ago. The positive aspects of this experience stemmed from one key personality trait...Curiosity. I was curious to learn about the 400,000 headlamps and tail lamps I was now responsible for buying, supporting engineering changes for, and ensuring the plant did not shut down if not supplied with parts on time. Aside from an Engineering Technology degree from Texas A & M, and 3 years as a Diving Officer in the Pacific, I had no experience of applying to this complex role.

Therefore, I made it a point to spend much time with engineers, visit our assembly plants, and certainly to visit as many vendor production plants as possible. With each engineering related interaction and exposure, I asked as many questions as possible and compared that to the small percentage of knowledge I gained from each vendor production facility visit. Why did one vendor organize their plant one-way verses another? Why did the engineer select this grade of blow molded plastic, or allow only a certain amount of reground (recycled) plastic in the production process? Why did certain headlamps and tail lamps get fogged with moisture over time? What "should" a headlamp or tail lamp cost? What was the average cost of a quality tail lamp? And certainly, how did Engineers view Purchasing...were we considered a bunch of contract writers to be brought in once the engineer decided who the supplier was going to be? Alternatively, were we viewed as driving the *best practice of having Engineering define the Clear (technical) Requirements, and Purchasing then engaging suppliers with RFI/RFPs and then validating with Engineering and doing Value Analysis before final decisions and business was awarded applied?*

If one reads this section and sees more questions than answers…that is EXACTLY the point. CURIOSITY, beyond many other skillsets, including analytics, strategic thinking, and emotional intelligence, is one of the most important skills a Buyer, Category Lead, Purchasing Manager, and certainly a Director of Sourcing or Supply Chain should have and continue to maintain.

Think of this. How many times have you walked out of a briefing, technical or otherwise, and not understood the primary goal of the discussion? Compare this with being in a meeting and being the one who respectfully asked that "dumb question" which, once answered, clarified the goal or issue for everyone in the room. Has anyone ever pulled you aside after the meeting and thanked you for asking that question? Be that person, not only to improve your knowledge, but to add value to the purchasing function and certainly the collaboration between Engineering and Purchasing.

Yet another benefit of understanding how Engineering views Purchasing and vice versa is from the perspective of endgame. At times, Engineering views Purchasing as simply a means to get the design they have for a keyboard, semiconductor chip, missile fin, or sportscar instrument cluster into production once the paperwork (contract) is inked. On the other hand, Purchasing may view Engineering simply as the team that produces a set of drawings which will be part of a bid package, which may lead to a chunk of their annual savings target. Both of these perspectives may be accurate, but there is so much more that can be accomplished for both shared services in the bigger picture.

Some in the Supply Chain community shy away from including additional functional representatives in negotiation meetings, since this may cloud the focus of commercial discussions, or even negatively impact the negotiation overall. I generally disagree. With preparation, cross pollination, and application of the vital concept of "One Voice" (from my previous book, Control what you Can) when several functional contributors collaborate, the results can be exponentially improved for all the shared services involved. One of the most effective forms of preparation includes role play, ideally with sales representatives from your own organization. Inclusion of unrehearsed scenarios, issues, and challenges from both sides improves the skillsets and agility of both Purchasing and Sales in negotiation situations and beyond. This brings to mind a quote from one of the best-selling books of all time… "As iron sharpens iron, so one person sharpens another."

Building on the idea that *Curiosity* is perhaps the best quality in a Category Leader, I have always found it interesting and beneficial to understand as much about the product and product design of the parts I am sourcing. As an engineer by degree, I can "speak" engineering, but will never be able to design complex parts, nor do I wish to. But, when I take the time to share my sourcing goals with engineers, while understanding why the engineer designed the part and his/her goals around this design, many more possibilities emerge. Once these things are mutually understood, and there are also mutually set boundaries, we can speak with One Voice to the supply base.

I love it when Engineering and Purchasing are aligned. If a vendor states that there are no further options to slim down pricing, but our engineer surfaces an opportunity based on design, this is a win not only between Engineering and Purchasing, but also a win for the vendor. Note of caution…even in circumstances where a sole source may be required due to only one vendor's ability to design a part or vendor having intellectual property rights mandates one vendor selection, it is vital that the engineer NEVER surface this in separate design focused discussions and certainly NOT during a cross-functional negotiation. Nothing has the potential to result in a cost increase than a perceived lack of competition!

Perspective with Purpose

Never stop being curious...to hone your trade and foster clarity/collaboration.

AI Perspective

Engineering teams are more routinely consulting AI for design solutions, including case studies applied to use of digital twins. The resulting identification of operational root causes may also lead to opportunities to deliver value and savings.

7

SUPPLIER QUALITY

&

THE TRIAD

Following the benefits I learned from *curiosity* noted in the previous chapter, the next step up in my Procurement and Sourcing education came in the form of The Triad.

Successful collaboration between Engineering and Purchasing is not a given and cannot be overrated. When it does not occur, the result is typically a lack of significant understanding, leading to wasted time, money, and effort. Unfortunately, this also tends to lead to counterproductive functional friction in the form of escalating conflict. Perhaps the most significant benefit of my engineering degree at Texas A & M was the ability to "speak engineering" and understand the engineering mindset. Linked to chapter 1, I have always done my best to foster the best relationships possible with my engineering counterparts and have certainly learned much as a result.

The most effective mini cross-functional teams I had the privilege of being part of were spurred by the concept that the combination of expertise which stems from commercial, technical, and quality parameters is vital. As noted above, I was benefiting from collaborating with my engineering contacts, but there was another, more significant factor yet to be applied in my first role at a major automotive supplier. As part of the planning for a new car model, engineers brought in purchasing once the design was mostly finalized to discuss design parameters and start to engage potential suppliers. The best mini cross-functional teams also included the Supplier Quality Manager (SQM). His/her responsibility is to ensure that the technical and material specifications are met, conduct supplier audits, and identify related issues. In addition, they work with vendors, engineering, and purchasing to identify the root cause of problems and implement the recommended resolution. This could include manufacturing process audits.

After the initial set of planning meetings with engineering and supplier quality, as the purchasing lead, I issued RFIs to a short list of suppliers and after joint evaluation of results, I planned visits to vendors. The benefit that this Triad of engineering, purchasing and supplier quality had on the process applied before, during and after-site visits and cannot be stressed enough. Much of this is due to one function surfacing an issue that influences or is balanced by, or can be resolved by one or two of the other functions. For example, a supplier may note that the grade of plastic that engineering has specified is the cost driver causing the price of the part to over-run budget. Based on a previous supplier audit, Supplier Quality could recommend adjusting the heat level when molding the part or propose using another grade of plastic.

I can attest that on the best Triads I was part of each function gained much insight from the other...and this was applied to the next design project, even before it started.

Perspective with Purpose

Prioritize the Triad of Engineering, Purchasing, and Supplier Quality to seek opportunities to deliver operational value and potential savings.

AI Perspective

AI prompts which include complex Engineering/Purchasing/Supplier Quality parameters may avoid a single solution that negatively impacts another parameter.

8

DATA ANALYTICS

&

PICK UP THE WENCH

My colleague & friend, Dr. Gulshan Singh, who manages a global team of analysts and related training, once shared the following analogy. **WE** help organize data and train associates on how to access it, **YOU** are then expected to "Pick up the Wrench, experiment, and learn how to use it."

In some larger organizations, Data Analytics teams take requests for data and related goals. They then not only identify the data, but generate relevant reports, and may even make recommendations as part of the delivered product. This is fine if your organization can support this, and certainly there are very talented analysts & data team managers who excel in this area. But I believe the norm in the age of information and flatter organizations is shifting in another direction. This direction involves expectation that individuals in any function become proficient in data mining, use of advanced data tools, and certainly the ability to set a course and develop a Sourcing Strategy based on where the data is pointing. This significantly increases the speed of several aspects of business, including Data Driven Decisions. It also helps to justify your role, along with that of the Supply Chain function, delivering savings and keeping the enterprise competitive.

AI is and will continue to play a central role in keeping corporations competitive. In a recent meeting I attended with a group of Chief Procurement Officers, it was brought to light that very shortly leaders will not only be expected to manage people but also AI agents that they are expected to ID to fill tactical gaps and create. This pivotal tool can certainly also be used for data modeling, including digital twins of sites, organizations, and software tools. But the responsibility lies with the leader to validate the data to avoid costly mistakes due to hallucinations.

Another basic but often forgotten formula for success & strategy as it applies to data, includes both iterative improvement and forward momentum.

First, the iterative improvement. Speaking from experience, seldom do learners soak up a new process, or in this case a new data tool immediately after training. Per the example above, one needs to first take the initiative to Pick up the Wrench…then accidentally drop it, apply it incorrectly, or most importantly try a few times or even *try terribly.* The idea is to learn just a little bit more each time. We all know this, but many hesitate to even achieve that 1% improvement which is the starting point to eventual mastery.

The second goal is the forward momentum applied to the data. This is more than the idea of Analysis Paralysis vs. Shoot from the Hip. My father on many occasions reminded me to "Make the best decision you can, with the best data you have at the time." This perspective not only provides forward momentum but also makes sense when looking back on a mistake, or disaster as the case may be. In most of the global corporations across several sectors I have worked for, before the age of the internet, and certainly before the current age of AI…the data set has never been, and will never be, COMPLETE. Leaders make the best use of the (data) tools available, consult with Subject Matter Experts, map a proposed strategy, discuss the merits and risks with their leadership, gain alignment and refinement, and Move Forward. The strategy can be adjusted along the way by checking in with stakeholders and leadership but those who are focused on delivering value plot a course and execute.

Perspective with Purpose

Pick up the Data Wrench, & make the best data driven decision possible, even if your data picture is not complete.

AI Perspective

Data modeling is one of the best applications for AI for experimentation. But leaders must ensure: 1) To maintain a clean benchmark set of data and 2) To check any AI developed solutions against sensible solutions based on your knowledge of your category to validate that solutions are hallucination free.

9

PURCHASING OPERATIONS

&

SUPPLY CHAIN SUPPORT

During my first role as a buyer at a major global automotive manufacturer, the day of the week I looked forward to the most was Friday...after I made it to 4:00 p.m..... That's because I could bet with 80 percent certainty that close to 3:00 on Friday I would receive THAT call. The call that would determine how the last few hours of Friday and the weekend would unfold, and how my manager would greet me on Monday morning. The call typically came from a manager on the assembly line in one of our US assembly plants with the news of a part shortage which would most likely shut down production…if I did not escalate the situation immediately to senior management at a key supplier. Due to this request, I would cancel all meetings for the day and the first part of the following week to solve this critical issue, since every minute the assembly line was stopped cost the corporation millions of dollars in lost production.

I can confirm that a year or so later, an experienced Purchasing Operations team was formed to handle these issues and going forward clearer scopes were established between Sourcing (Category Leaders) and Procurement (Tactical Buyers). Let's explore industry standard advantages of each, and just as importantly, how they complement each other.

Strategic Sourcing / Category Leaders

Before diving into each of these sub-functions, it is important to understand that all are vital to the Supply Chain/Purchasing shared service function to deliver value. Sourcing cannot function unless Procurement and Operations execute their roles, and likewise Procurement cannot engage at a tactical level without clear strategic direction in the form of comprehensive contracts. Certainly, Operations needs not only guidance from Strategic Sourcing and Procurement, but Strategic Sourcing must have input from both

Procurement and Operations to put in place effective contracts.

Category Leaders, aligned with industry standard practice, should focus their efforts on negotiating long-term contracts based on the longer-term Supply Chain strategy of the company. Drivers for this can include new tech and new product platforms. Back in the early 1990's this included engagement & negotiation with a variety of suppliers developing in-car navigation and infotainment. In the chemical sector this could include focus on innovative products related to electric vehicles. In the tech sector this might include the resurgence of wearable tech. In the construction sector, this might involve contracting in support of long-term, multi-billion-dollar plant upgrades to support the production that produces some of the innovative products above. Strategic Sourcing Category Leaders should be focused on these long-term contracts to implement the strategic vision of the corporation.

Procurement / Tactical Buyers

Tactical Procurement Buyers have the unique and vital role of executing & engaging with suppliers based on the strategy and contracts Strategic Sourcing puts in place. In order to be effective, they must be very aware of these strategies and Clear Requirements outlined in the contracts and certainly have solid working relationships with the Strategic Sourcing Category Leaders and consult with them often. Tactical procurement buyers also execute smaller scale/spend bids and contracts and will conduct more tactically focused Quarterly Business Reviews (QBRs) with their more critical suppliers based on input from vendors and KPI tracking. Some of this input may also originate from the Operational team. It is certainly appropriate for these buyers to participate in the Strategic Sourcing QBRs from time to time as well, especially for strategic and high-risk suppliers. Two points cannot be overstated.

1) The necessity for two-way communication and the resulting collaboration between Strategic Sourcing and Tactical Procurement.

2) The concept that one role is not more important than the other as well as the importance of boundaries between the two.

Category Leaders must not spend excessive time on tactical matters, or they will lose sight of the strategic horizon, and Tactical Buyers must stay focused on day-to-day management of suppliers and smaller bids to ensure continuous supplier support. But it is also key that Category Leaders need to maintain basic level of awareness regarding tactical tasking and related vendor KPIs which apply to strategic sourcing. A similar reverse perspective is useful for Tactical Buyers relative to understanding strategic goals linked to global initiatives.

Purchasing Operations / Back Office Support

Many multinational corporations establish back-office support centers for a cross-section of shared services, including Purchasing/Supply Chain in countries with a comparatively lower labor cost within a particular region. In many cases these regional or global business centers are housed in a business complex which may serve several other multinational corporations. In general, the advantage of lower cost labor outweighs the disadvantage of not having the (Purchasing) Operations team housed within the headquarters facility. However, keep in mind that these employees (sometimes contingent workers) could be inclined to shift between one multinational and another for small differences in salary or working conditions, since their commute will not differ and they may be networking closely with employees on the next floor or building.

When considering the collaboration between the Purchasing Operations team and Tactical Buyers, key points to consider are the Clear Requirements and role definition. As outlined in the anecdote in the first part of this chapter, in some corporations, the role of the

Operations team is to proactively manage tasks related to the "Pay" sub function of the Source to Pay shared service. Included are invoice processing and resolution, Travel and Expense (T&E) card processing and issues, P-Card processing and issues, and other tactical process-related tasking. Again, this group of buyers is no less or more vital to the overall Purchasing/Supply Chain shared service, and it is extremely important that feedback is given to the Tactical Procurement Buyers related to which vendors are not performing to the established set of KPIs, so that when support in the form of escalation is needed from the Tactical Procurement group, (and if needed Strategic Sourcing), it can be applied. It is critical that the entire shared service speaks with One Voice to a global vendor.

Certain multinationals may choose to combine the Tactical Buying team with the Operations team based on budget or resource constraints. However, based on industry standards, establishing three separate groups, ideally with the Strategic Sourcing and Tactical Procurement teams located in the regional headquarters is best practice based on synergy and Total Cost of Ownership. A key reason for this is the efficiency and necessity for these two teams to engage in person with their vendor counterparts in the home country, either at HQ or at the vendor's HQ or production facilities.

Typically, the Purchasing Operations team reports up through the Purchasing/Supply Chain Business Process team, responsible for establishing, enforcing, and monitoring processes and systems that support the overall shared service. A note of caution should be considered relating to decisions around business processes and related systems. As the Purchasing/Supply Chain function is a support function to the overall enterprise, so too is the Purchasing Operations/Business process subfunction to Purchasing/Supply Chain. It is wise to maintain this insight when strategic decisions are being made to optimize the value delivered at the strategic level of the overall function and leadership of this sub function. In simple terms, avoid the tail wagging the dog. That is, Strategic Sourcing/Supply Chain should lead by identifying any process or system gaps that need to be addressed and assign this to the Business Process/Purchasing Operations team to identify and propose solutions. These proposals should be vetted by the Strategic and Tactical buyers (Stakeholders) before approval and implementation. Aligned with this concept, it is advantageous to have senior leadership of this sub function comprised of at least one member with significant Strategic Sourcing experience, and another with back office/Purchasing Operations experience.

Perspective with Purpose

Tactical and Operations support of Strategic Sourcing, along with regular feedback, and continuous improvement, are vital to value delivery.

AI Perspective

AI's most useful application should target defined, repeatable tasks. These tasks are focused on operational tasking and will result in significant process improvement and value delivery.

10

SITE

&

BALANCED PERSPECTIVE

When an assembly plant, warehouse, site, or rig operational leader responsible for the day-to-day production of product reaches out to the Supply Chain/Procurement team with a supplier related issue, it's important. Why? Because this is how your corporation makes a profit and stays in business. Their perspective matters in a big way. This is where the rubber meets the road, whether your enterprise produces laptops, chemicals, cars, furniture, or oil.

Why then, over 30+ years of my Supply Chain career across Automotive, Logistics, Oil & Gas, Construction and Chemicals, has there been so much consistent friction and misalignment between those responsible for managing plant production and those responsible for purchasing the goods and services to support the plant?

In my first book, *Control what you Can*, the foundational chapter was centered around the idea of CLEAR REQUIREMENTS. This is a phrase uttered often in our office, either as a reminder when considering next steps, or in retrospect when conducting a root cause analysis. The foundational catch phrase for this book is **BALANCED PERSPECTIVE**. It will result in exponential improvement in purchasing and resulting production. Let's look at both sides of this equation.

Production Perspective

I've supported supply chain organizations where plant leadership had ultimate control of every aspect of what was purchased, how much, at what price, which suppliers, how often, etc. Why wouldn't this be the case? As noted above, producing the maximum amount of required product when it is needed is the ultimate goal of any for-profit enterprise. This structure in most cases includes site-based procurement teams that procure required material based on guidance from those at the site (or inventory control) who are forecasting demand. It is typically very Agile, and likely involves one production site, or if more sites are part of the enterprise, they have their own local procurement teams. These buyers are often nimble "jacks of all trades," who are skilled at buying all kinds of goods and services, in many cases on an as needed basis if emergencies pop up. Relationships with suppliers are an absolute priority, because there is not a lot of time to spend on administrative paperwork, including lengthy contracts with extensive terms and conditions. There is also little time for a bureaucratic and lengthy approval process, which in many cases delays securing needed parts/raw materials to produce product and generate profit.

Supply Chain Perspective

I have also supported supply chain organizations where the corporate HQ based supply chain had ultimate control over the elements noted above, with leadership focused on more strategic elements including consolidating the supply base, leveraging volume, ensuring contracts were legally air tight, and in favor of the enterprise, with a robust approval process driving accountability which would meet the scrutiny of internal or external auditors. Why wouldn't any enterprise want these elements in place to support the competitive purchase of goods and services? Besides, any knowledgeable and strategic sourcing practitioner knows that every Dollar/Euro/Peso/Yuan/Real, *saved* goes right to the bottom line, with no Selling, General & Administrative expenses subtracted. With a reliable forecast, delivered well in advance, there should be no significant issues to supporting the plant. And certainly, when part designs, project scope and manning requirements are delivered with a reasonable amount of time to work through a typical seven-step strategic sourcing process, the best strategy can be developed, implemented, and adjusted for maximum efficiency. When early engagement by the plant or, at least, *reasonable engagement* timelines are practiced, a centralized supply chain organization can execute and deliver.

Balanced Perspective

Now let's explore plant and supply chain reality. You know it, and you have lived it. The two extreme perspectives outlined above mix like oil and water if senior leadership is siloed. Organizational design is fundamental when attempting to understand why some enterprises have little or no friction between plant and supply chain, while others are their own worst enemy. The organizations I have worked for that have come closest to the first instance typically have representation from Purchasing in the C-Suite in the form of a Chief

Procurement Officer (CPO) or at least report to the Chief Financial Officer (CFO) or the Chief Operations Officer (COO). This industry standard model is ideal since it balances the footing of Supply Chain with that of other shared services. It supports constructive engagement when these functions may have different perspectives related to significant challenges. These could include the perceptions noted above relating to site leadership and centralized Supply Chain, or the balance between Engineering and Supply Chain related to supplier selection. Based on engagement and escalation initiatives I have led at various levels in several business sectors, the tone is set from the top of the organization and permeates down to the individual contributor (buyer, engineer, project manager, etc.) The organizational design, related tone, and *Balanced Perspective* can either support needless escalation and debate, or positive collaboration and resulting wins in the marketplace.

So, when you receive that call from a frustrated project manager at your assembly plant/production site who is seeking help regarding a challenging supplier situation, first consider their perspective and their focus on producing product. Then *balance* this with your consolidation initiative. But first listen, seek to understand your ultimate stakeholder, and map a solution together with ideally a short-term solution to get through the immediate crisis. Then, commit together to work toward a longer-term *balanced perspective* solution that will remedy the situation going forward. The concept of stakeholder management is certainly vitally applicable when considering options to apply AI. It is key to ensure that stakeholders are aware and consulted regarding application of AI agents or digital twins linked to sites or vendors (as noted in chapter 1).

Perspective with Purpose

A Balanced Perspective is vital when bridging the gap between assembly plant stakeholder immediate needs and strategic supply chain goals.

AI Perspective

While AI is very effective in modeling demand planning, it must be paired with consistent stakeholder engagement to ensure overall plant/supply chain collaboration.

11

CHANGE MANAGEMENT

&

THE CRISIS

Recently while delivering a Strategic Sourcing class internally to a group of new category leaders along with leaders from few other shared services and those from sites across the Americas, we engaged in an especially lively discussion as part of the change management framework section. The perspectives and related behaviors reflected factors that either built a solid foundation for value delivery through positive process change, or "essential watch outs" that are just as valuable to steer clear of, when embarking to implement a Strategic Sourcing program.

As with a basic sourcing exercise, before diving into these factors, let's establish a baseline for change management. There are several generally accepted models, and most involve a few basic stages. They include:

- Being cognizant that a change needs to be made.
- Wanting to make the change.
- Providing the group with the basic information to make a change in behavior.
- Giving the team the capability/skills to make the change.
- Reinforcing the change, through repetition and established goals.

I have led and been on the receiving end of several strategic sourcing change management initiatives, and from research and personal experience have found that those that are successful are based on several identifiable success factors and overcome, or at least mitigate, a few very common obstacles. As with most challenges and negotiations, it is extremely beneficial to look at these factors from a few different perspectives.

Based on my Navy Diver training which required constant vigilance around what could go wrong, let's consider obstacles and related perspectives. Why would buyers in an organization without category management or strategic sourcing resist change? Consider the perspective of a buyer who has been rewarded for completing contracts and related purchase orders after a business or site leader had selected the supplier, price, and had initiated the "paperwork" requests to achieve an operational goal at a production site. This buyer was compensated based on the speed of processing this paperwork and occasionally sending out three bids and a buy. Consider the perspective of the site leader who sent these paperwork requests to the buyer over the years, who expecting nothing but a quick response and saw no reason to consider the advantage of combining volume across sites to improve cost and service. From his/her focus on solely optimizing production at one particular site, this would only cause delay. Consider also the management structure above these buyers. It is easy to focus on the volume of paperwork and processing speed. All of the above are very common reasons for buyers, purchasing managers, and certainly business and site leaders to resist implementing even a basic category management program, let alone a full-blown strategic sourcing program, especially if the business enterprise is experiencing reasonable profitability.

Most readers are very familiar with the advantages of a standard strategic sourcing program anchored by category management. They include maximizing value by categorization spend, buyer expertise, leveraging volume across sites, regions and globally, establishing vendor KPIs related to a Supplier Relationship Management (SRM) program, and several other factors which result in.

- Assurance of *Supply*
- Improved *Service*
- More reliable *Quality*
- Competitive *Price*
- Best *Innovation*, which improves enterprise competitiveness

Now let's consider the perspective of the director hired as a change agent to implement & drive a strategic sourcing program. Having been in this role on more than a few occasions, I can prepare the reader with a few expectations when facing the obstacles noted above, when seeking these advantages…and the levers and attitudes which may be helpful.

As outlined above, buyers at less mature purchasing organizations are not expected to guide, train, and positively influence stakeholders and have little incentive to do so. This is before one considers the factors outlined at the beginning of this chapter to drive and implement change. As a change agent, one will have to understand very clearly, the fundamental reason business and site leaders will resist change...the perspective of **LOSS OF CONTROL**. This is difficult for any individual contributor or leader when business is going well and even more when the business faces challenges.

This is why one of the initial elements of any change management program and especially one focused on the shift to strategic sourcing is the unwavering support of senior management. At a minimum, this should be secured and driven by the Chief Procurement Officer (CPO) or lead of the shared service and, if possible, the CEO. This further reinforces the change, as it will impact leaders and individual contributors from other shared services and business units. It is also best to set a realistic timeframe with milestones for implementation (including ROI targets) of this significant change. If the enterprise currently has little focus on strategic sourcing and is of reasonable size (8,000-10,000 employees), a five-year timeframe is realistic.

Most strategic sourcing programs are centered around a repeatable process. A typical process may include 7 steps which could include:

- Identifying opportunities
- Internal analysis
- External analysis
- Strategic options
- Go-to-Market Strategies
- Implementing Strategies
- Continuous Improvement

There are several versions of these programs, and some may include as few as four steps and others may break this down into as many as twelve or more. There are books and articles which engage in these steps in great detail, but we will only outline this version at a high level as the focus of this chapter remains the *perspectives* involved when implementing a strategic sourcing program.

Highlighting opportunities involves generating high level improvement ideas based on basic knowledge of the category. *Internal Analysis* delves deeper into spend analysis, stakeholder review, and needs of the business. *External Analysis* focuses on review of market intelligence outside of the business, along with comparative pricing, what a product should cost, and Total Cost of Ownership. The *Strategic Option* step centers around which sourcing strategy may be most effective. The *Go-to-Market Strategy* step involves review and implementation of sourcing strategies. *Implementing Strategies* is the follow-up after sourcing to ensure initial supplier delivery of the quality, service, technology, and price, along with other sourcing goals aligned with stakeholder needs. *Continuous Improvement* focuses on monitoring Key Performance Indicators (KPIs) that should be part of the contract...which includes the CLEAR REQUIREMENTS, as noted in my first book, **Control what you Can.** It also focuses on working with the supplier to deliver additional step changes in value for the company.

It is vital to remember that this process is NOT linear and is never complete. To be sure, once the last step of this process is complete, the process should be initiated again. (The best depiction of this process is a wheel.) The timing for a smaller category with lower spend or significance, could be a few weeks, and for a larger more significant category, it could be 3-5 years. Thus, documenting these steps for the current category lead, and those who will follow, is of great importance. As with any process, it is highly recommended to document, by way of a charter, the overall process itself.

This process should also include a regular meeting of senior leaders from Purchasing, other key functions typically including Finance, Legal, and as applicable, Supplier Quality, Engineering, and others. This drives the most critical element of the process…. STAKEHOLDER ALIGNMENT. While the Category Lead is expected to develop the category strategy based on the steps above, he/she is expected to validate this proposed strategy with business and site stakeholders as it is being developed. TCO, or Total Cost of Ownership, should be included in any category strategy and related planning with stakeholders. In some cases, the strategy will need to be tweaked, in others significantly adjusted, and at times drastically changed. This may be due to the priorities of these stakeholders, data issues, or changes in foundational elements, including budget and geo-political factors (wars, tariffs, etc.). To be sure, when the Category Lead first initiates the strategy with his/her mid-level counterpart, it is not a "tell" but a recommendation, with sincere request for validation or targeted improvement.

Best practice in the application of a new process typically includes a RACI chart. (RACI = Responsible, Accountable, Consult, Inform). It is vital to note that only one party be listed as Accountable.

Consider the case of a church that loses the majority of its members and reputation when the charismatic pastor that the congregation admires departs. As with any successful change management program, it is highly recommended to center the change and success of the program around the process, and not a few key individuals or change agents. This will ensure that if the change agents depart, the positive changes and momentum instilled remain in place across the global enterprise. As noted above, standard practice would include setting up a charter to document the process and related KPIs along with the success of savings and other achievements.

Remember the last time your enterprise faced significant challenges in the marketplace? This could be due to falling profits, new competition, or other significant crises (Remember COVID?). Supply Chain leadership, along with most other leaders, has two alternatives. Simply keep executing where possible and hang on to see what happens…. or TAKE ADVANTAGE OF THE CRISIS. In most financial crisis situations, Finance and other leaders turn to Supply Chain/Purchasing to attack the supply base with basic savings initiatives…(ever been asked to call a supplier in and blatantly ask for 10, 15, 20+% savings?) Instead, why not outline a detailed plan based on the Strategic Sourcing steps above, with reasonable goals, and seek approval to apply this to that sacred cow that in the good times was out of scope, because they were "printing money". And when, not if, significant value is delivered, be sure to secure top management support to ensure the change management process your team worked very hard to implement is clearly documented and approved by senior management…when the good times roll again. It is also wise to focus on the issue, not the personality or group, when engaging in a continuous improvement initiative. A perspective that includes a rational, professional approach will result in less pushback and significantly more acceptance, ownership, and overall benefit to the organization.

Perspective with Purpose

Take advantage of a good crisis and implement a Strategic Sourcing process via an effective change management initiative.

AI Perspective

Apply AI in as many of the strategic sourcing steps as possible to maintain focus on the higher-level strategy, especially in the realm of internal and spend analysis.

12

PERSPECTIVES WITH PURPOSE

&

LEGACY

While this book has focused on the perspective that Supply Chain/Purchasing leaders have regarding other functions, these functions certainly have a perspective when it comes to our function.

After a lengthy questionnaire and perhaps an interview, most personality-related leadership training then buckets employees into 4+ types (colors, animals, etc.). This is followed by training which explains the virtues and pitfalls related to which bucket(s) the employee falls into…and in what situations the employee may lean into. This is focused on a deeper understanding of your leadership and engagement style…and outward looking perspective.

It is VITAL to remember, however, that those, for instance in Engineering, are likely to have a perspective on Supply Chain based on their past experience with Supply Chain. When invited to speak to university students regarding my previous book, *Control what you Can*, I make it a point to highlight the idea that *"Your reputation and that of purchasing arrives at the table, before you sit down."* With this in mind, if you happen to be new to your company, role, or Supply Chain, it would be extremely beneficial to objectively and discreetly investigate the current perception of your function. Whether positive or negative, this will immensely assist in helping you to understand what you are stepping into. Applying my Navy parlance, do you need to approach your new assignment as "All Hands-on Deck" (building on solid reputation with initiative), "Steady as she goes" (fix a few issues, but keep going in same strategic direction), or "Damage Control" (prepare for much push back, deliver quick wins)?

Zooming out yet further, consider your overall perspective on your Supply Chain career. What are you striving for…or put another way, what kind of leadership legacy will you leave?

I can count on one hand the model leaders who have delivered significant savings, maintained their technical edge, but above all, have been models of **Balanced Perspective.** These leaders made mistakes along the way, took responsibility, and did their best to teach and mentor others even under less than perfect circumstances...while considering a variety of perspectives. They demonstrated the same level of courage, trust, tough but fair expectations, technical expertise, and honor...in front of large audiences but, even more importantly, during one on one sessions with the door closed.

These leaders also modeled another related key leadership trait that is beneficial not only in a business setting, but perhaps just as life elevating, with personal relationships. From a self-realization perspective, they modeled the trait of responsiveness when receiving criticism. This is not easy for most of us. It is easier to respond to and internalize feedback when delivered in a respectful manner from one we trust, but certainly more difficult to take onboard when not received from your circle of trust.

How do you respond when called out bluntly, or even given constructive criticism in a gentle way, by a colleague from another shared service? Most of us are challenged to say the least. A recent study indicates that when even slightly negative email input is received from someone we do not know, most perceive it as an attack, and categorize it one or two levels more significantly negative than if we received it from someone we trust and who has our best interest in mind. These reminders are crucial when applying a *Balanced Perspective*:

- Wait an hour or perhaps a day before responding
- Seek first to understand the input

Certainly, do your best to follow through with what is best for the enterprise even if this means admitting a mistake. Think about this. How do you perceive someone who sincerely takes responsibility for a mistake, internalizes what happened, and puts measures in place to limit the likelihood of the same occurrence? Most of us are surprised and elevate our perspective of this person a few more levels up the respect ladder. Also consider the root of the word mistake…this occurs when one MIS-TAKES, or mis-understands the situation, and therefore applies a different perspective when selecting from their knowledge base and how they plan to resolve the matter. We have all done it. And it would improve our value delivery and stress level by applying more grace in these situations.

Think of how actions, decisions, and communication patterns you demonstrate influence how you are perceived in the workplace. Is this how you want to be perceived? Is this your intent? The perception and the perspectives others have of you *is their reality*. Are you making decisions, as expected by leadership of the corporation, that ultimately deliver value to the corporation, rather than to you as an individual contributor, leader, or the Supply Chain function? A good litmus test is engagement with a trusted internal mentor, who will surface candid feedback during regular catch-up sessions. Even when it is tough to take onboard at first, these truths regarding other's perceptions of your leadership engagements will forge you into the leader you strive to be. While everyone wants to be perceived as one who "gets things done," this is not effective in the long run if this results in emotional casualties. A 360 review process administered by HR can also be helpful, but even according to HR leadership, results should be tempered by the reality that at times both reporting staff and those from other shared services may skew results. Since responses to these surveys can be utilized as venting opportunities...originating from frustration with leadership, cumbersome processes, or manning levels.

When driving towards a Balanced Perspective, consider the postures below that can positively impact the perspective of others when leading a cross-functional team, as many Category Leaders are expected to do on a regular basis:

- Assume positive intent
- Focus on the problem & process first (not the person)
- Listen more, talk less
- Give credit more than seeking ways to take it
- Admit fault, ID root cause, install fix, move on
- Realize there is another perspective, and consider it

What is senior leadership seeking?

I recently had a discussion with senior Supply Chain leaders from major global High Tech and Oil & Gas corporations. It was centered on the capabilities expected of the next generation of leaders in the current dynamic business climate. They confirmed that leaders with perspectives which kept them focused on understanding the problem at hand from various shared service and business angles was vital. This, combined with the ability to optimize and deliver streamlined solutions, allows corporations to outmaneuver their competitors in our new, rapidly evolving global marketplace. These are key traits corporations are seeking in the next generation of leaders. Put simply, this new leader will not be of the standard mold or from one specific shared service (Supply Chain, Engineering, Accounting, etc.) sought after in recent years. This leader will most likely be a hybrid leader who applies a **Balanced Perspective** on a regular basis.

As an Eagle Scout, and later coach of Eagle candidates, I shared with these fledgling leaders in their final review board this key concept: You will learn from every leader you have. From poor leaders, you will hopefully learn where not to imprint patterns (self-serving, arrogance, blame finding, etc.). From admired leaders you will find inspiration (competence, servant leadership, courage, character). From Control what you Can, we learned that each day, your Mindset, or how you Set your Mind, is vital. How you approach each decision/situation of each day results over the years in the legacy you will leave, not only throughout your career, but your life outside of work. As with your faith life, you are either seeking and getting closer (but never reaching) that perfect state of leadership and refined faith...or wishing you had. If one approaches these decisions/situations with courage, selfless engagement and commitment, the trend will be an upward trajectory. Things will not always be easy, but moving in a positive direction, resulting in a legacy you will be proud to leave others.

At a local, national, and global level, if we apply a *Balanced Perspective* to not only in the workplace but to individual interactions, no matter which side of an issue we may find ourselves, we might be able to *Control what we Can* to minimize much of the anxiety and geopolitical strife in the world today.

Perspective with Purpose

With each decision, leave a Legacy of Leadership others will be proud to follow.

AI Perspective

Applicable to this and previous chapters, AI is extremely effective in developing time saving solutions to repeatable tasks, potential solutions, data/information gathering, and modeling options.

Ultimate responsibility, and strategy should always remain with the Human Leader.

PERSPECTIVES WITH PURPOSE

P – W – P

ACKNOWLEDGEMENTS

None of the creative content of this book was derived from Artificial Intelligence (AI)

I would like to thank Jeff Swift, Tom Derry, Jeff Martin, Adrian Bregnard, Jim Guinn II, Xenophon Koufteros, Achim Heyne, John Ploetz, Dr. Gulshan Singh, and Lisa Haley for their encouragement and honest feedback. And Tim Shulte and *Variance Author Services.*

Special thanks to Rick Kovacich and Jose Manuel Velarde for their trust, guidance, and friendship as we led the Americas Huntsman Indirect team. It has been a rare honor and privilege.

And God, who blessed me with this life, family, country, encouraging friends, leaders, experiences, and abilities that allow me to serve Him.

LAUS DEO

ABOUT THE AUTHOR

Peter Dill has led global Supply Chain teams for 30 years (including 16 years overseas) for the U.S. Navy, General Motors, CEVA Global Logistics, FMC Technologies, and Huntsman Corporation. He earned a Bachelor of Science in Engineering Technology from Texas A & M University, where he was a member of the Corps of Cadets, as well as an International MBA from Thunderbird. He is a Past President of the Houston chapter of the Institute for Supply Management(ISM), where he currently serves on the Executive Board.